READING
ABOUT SCIENCE
Skills and Concepts

BOOK C

John Mongillo
Beth Atwood
Kevin M. Carr
Linda J. Carr
Claudia Cornett
Jackie Harris
Josepha Sherman
Vivian Zwaik

Phoenix Learning Resources

ISBN 0-7915-2203-2

2 3 4 5 6 7 8 9 00 05

Authors

John Mongillo, Senior Author and General Editor
Science Writer and Editor
Saunderstown, Rhode Island

Beth S. Atwood
Writer and Reading Consultant
Durham, Connecticut

Kevin M. Carr
Teacher and Writer
Honolulu, Hawaii

Linda J. Carr
Writer and Psychologist
Honolulu, Hawaii

Claudia Cornett
Professor Emerita
Wittenberg University

Jackie Harris
Medical and Science Editor
Wethersfield, Connecticut

Josepha Sherman
Writer and Science Editor
Riverdale, New York

Vivian Zwaik
Writer and Educational Consultant
Wayne, New Jersey

CONTENTS

Do you enjoy the world around you? Do you ever wonder why clouds have so many different shapes and what keeps planes up in the air? Did you ever want to explore a cave or find out why volcanoes erupt or why the earth shakes? If you can answer yes to any of these questions, then you will enjoy reading about science.

The world of science is a world of observing, exploring, predicting, reading, experimenting, testing, and recording. It is a world of trying and failing and trying again until, at last, you succeed. In the world of science, there is always some exciting discovery to be made and something new to explore.

Four Areas of Science

READING ABOUT SCIENCE explores four areas of science: life science, earth-space science, physical science, and environmental science. Each book in this series contains a unit on each of the four areas.

Life science is the study of living things. Life scientists explore the world of plants, animals, and humans. Their goal is to find out how living things depend upon each other for survival and to observe how they live and interact in their environments, or surroundings.

Life science includes many specialized areas, such as botany, zoology, and ecology. *Botanists* work mainly with plants. *Zoologists* work mostly with animals. *Ecologists* are scientists who study the effects of air pollution, water pollution, and noise pollution on living things.

Earth-space science is the study of our Earth and other bodies in the solar system. Some earth-space scientists are *meteorologists*, who study climate and weather; *geologists*, who study the earth, the way it was formed, and its makeup, rocks and fossils, earthquakes, and volcanoes; *oceanographers*, who study currents, waves, and life in the oceans of the world; and *astronomers*, who study the solar system, including the sun and other stars, moons, and planets.

Physical science is the study of matter and energy. *Physicists* are physical scientists who explore topics such as matter, atoms, and nuclear energy. Other physical scientists study sound, magnetism, heat, light, electricity, water, and air. *Chemists* develop the substances used in medicine, clothing, food, and many other things.

Environmental science is the study of the forces and conditions that surround and influence all living and nonliving things. Environmental science involves all of the other sciences-life, earth-space, and physical.

If you want to know more about one or more of these areas of science, check the bibliography at the back of this book for suggested additional readings.

Steps to Follow

The suggestions that follow will help you use this book:

A. Study the photo or drawing that goes with the story. Read the title and the sentences that are printed in the sidebar next to each story. These are all clues to what the story is about.

B. Study the words for the story in the list of Words to Know at the beginning of each unit. You will find it easier to read the story if you understand the meanings of these words. Many times, you will find the meaning of the word right in the story.

When reading the story, look for clues to important words or ideas. Sometimes words or phrases are underlined. Pay special attention to these clues.

C. Read the story carefully. Think about what you are reading. Are any of the ideas in the story things that you have heard or read about before?

D. As you read, ask yourself questions. For example, "Why did the electricity go off?" "What caused the bears to turn green?" Many times, your questions are answered later in the story. Questioning helps you to understand what the author is saying. Asking questions also gets you ready for what comes next in the story.

E. Pay special attention to diagrams, charts, and other visual aids. They will often help you to understand the story better.

F. After you read the story slowly and carefully, you are ready to answer the questions on the questions page. If the book you have is part of a classroom set, you should write your answers in a special notebook or on paper that you can keep in a folder. Do not write in this book without your teacher's permission.

Put your name, the title of the story, and its page number on a sheet of paper. Read each question carefully. Record the question number and your answer on your answer paper.

The questions in this book check for the following kinds of comprehension, or understanding:

1. *Science vocabulary comprehension.* This kind of question asks you to remember the meaning of a word or term used in the story.

2. *Literal comprehension*. This kind of question asks you to remember certain facts that are given in the story. For example, the story might state that a snake was over 5 feet long. A literal question would ask you: "How long was the snake?"

3. *Interpretive comprehension*. This kind of question asks you to think about the story. To answer the question, you must decide what the author means, not what is said, or stated, in the story. For example, you may be asked what the main idea of the story is, what happened first, or what caused something to happen in the story.

4. *Applied comprehension*. This kind of question asks you to use what you have read to (1) solve a new problem, (2) interpret a chart or graph, or (3) put a certain topic under its correct heading, or category.

You should read each question carefully. You may go back to the story to help you find the answer. The questions are meant to help you learn how to read more carefully.

G. When you complete the questions page, turn it in to your teacher. Or, with your teacher's permission, check your answers against the answer key in the Teacher's Guide. If you made a mistake, find out what you did wrong. Practice answering that kind of question, and you will do better the next time.

H. Turn to the directions that tell you how to keep your Progress Charts. If you are not supposed to write in this book, you may make a copy of each chart to keep in your READING ABOUT SCIENCE folder or notebook. You may be surprised to see how well you can read science.

PRONUNCIATION GUIDE

Some words in this book may be unfamiliar to you and difficult for you to pronounce. These words are printed in italics. Then they are spelled according to the way they are said, or pronounced. This phonetic spelling appears in parentheses next to the words. The pronunciation guide below will help you say the words.

ă	pat	î	dear, deer, fierce,	p	pop	zh	garage, pleasure;
ā	aid, fey, pay		mere	r	roar		vision
â	air, care, wear	j	judge	s	miss, sauce, see	ə	about, silent
ä	father	k	cat, kick, pique	sh	dish, ship		pencil, lemon,
b	bib	l	lid, needle	t	tight		circus
ch	church	m	am, man, mum	th	path, thin	ər	butter
d	deed	n	no, sudden	*th*	bathe, this		
ĕ	pet, pleasure	ng	thing	ŭ	cut, rough		
ē	be, bee, easy,	ŏ	horrible, pot	û	circle, firm, heard,		
	leisure	ō	go, hoarse, row,		term, turn, urge,		
f	fast, fife, off,		toe		word		**STRESS**
	phase, rough	ô	alter, caught, for,	v	cave, valve, vine		Primary stress ′
g	gag		paw	w	with		**bi·ol′o·gy**
h	hat	oi	boy, noise, oil	y	yes		[bī ŏl′ejē]
hw	which	ou	cow, out	yōō	abuse, use		Secondary stress′
ĭ	pit	ŏŏ	took	z	rose, size,		**bi′o·log′i·cal**
ī	by, guy, pie	ōō	boot, fruit		xylophone, zebra		[bī′elŏj′ĭkel]

The key to pronunciation above is reprinted by permission from *The American Heritage School Dictionary* copyright © 1977, by Houghton Mifflin Company

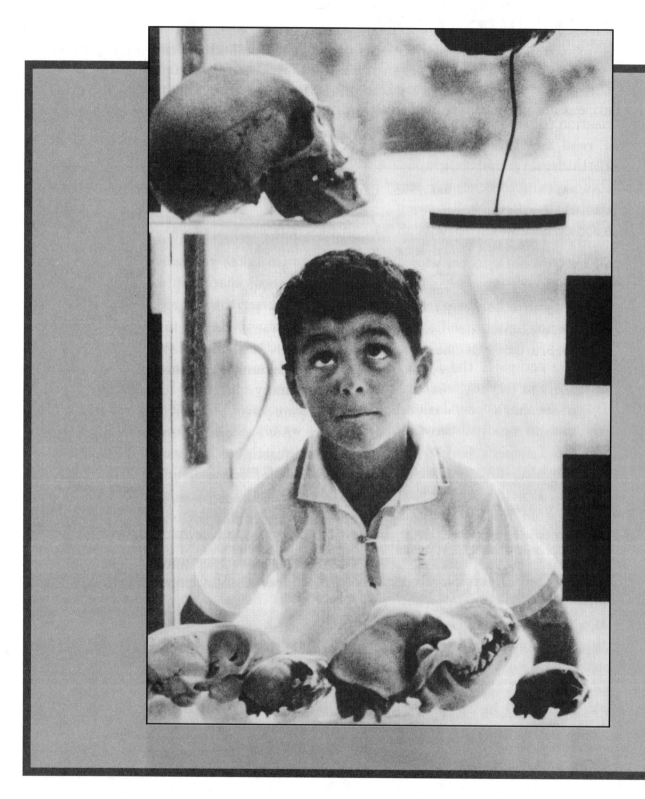

LIFE SCIENCE

A human skull has 22 bones. Of these 22, 8 bones cover and protect the brain. The other 14 are face bones. Only one bone in the skull can move; the others are joined tightly together. Do you know which skull bone is movable? Open your mouth. You moved your jawbone, the only movable bone in the skull.

WORDS TO KNOW

The Clock of Life
internal clock, "internal—inside of; "clock"—the thing that makes other things happen at a certain time
signal, message
chemical, a substance that causes something to happen
perform, to do, to act

The Big One
extinction, state of being out of existence
layers, several thicknesses
mammal, warm-blooded animal
plankton, microscopic plants and animals that float in bodies of water
pods, coverings for one or more seeds (a pea pod); a group of whales

Sharks!
cartilage, tough, flexible whitish body tissue forming part of the skeleton
elastic, being able to return to original size and shape
skeleton, all of the bones that form the framework of an animal or human being
torpedo, a large, cigar-shaped, self-propelled underwater bomb; something sleek and swift in the water

The Flashlight Fish
organs, body parts
conceal, hide

locate, find
reefs, a ridge of rock, coral, or sand near the surface of the water

The World's Largest Rodent
gnaw, to bite and wear away bit by bit
hind, back
webbed, having a membrane joining the "toes"

Sea Animal May Help Dentist
decayed, rotted
cavity, a hole, depression
attaches, to join, fasten to

Invite a Long Lifetime
organ, a body part
avoid, to keep away from
blood pressure, force pressing against the walls of the veins
excess, surplus, more than needed

Protecting Your Health
physical, of the body (as opposed to the mind)
properly, correctly
shots, injections

The Hardiest Weed of All
perennial, sprouts, comes, every year

Red Seaweed
product, something made
collect, to gather, harvest
ingredients, the things that are mixed together to make the product

Life in the Forest
energy, power
process, a series of actions by which
  something happens
survive, to live

Tree Rings
observe, to look at carefully
examining, inspecting, looking at to
find out the facts

Watch Out for Poisonous Plants
annoying, bothersome, irritating
recognize, to know on seeing

The Misunderstood Spider
misunderstood, not understand
  correctly
thorax, the middle of the three main
  parts of an insect

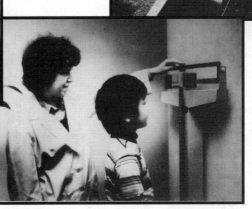

The Clock of Life

A chicken lays an egg. You feel sleepy. And a tree loses its leaves. All of these things, and many more, happen in a certain way, at a certain time each day or each year. They take place because of something called an *internal clock* (ĭn tûr′nəl klŏk′). The word *internal* means "inside of," and the internal clock is inside a certain part of every plant and animal. For example, there is an internal clock in the head of a chicken. The internal clock receives a *signal* (sĭg′nəl), or message, from the world around it. Some of these signals include light, heat, dark, and cold. When the internal clock gets the signal, the body of the plant or animal produces a chemical that causes the living thing to perform different actions. For example, daylight signals the chicken's internal clock to make a chemical. Then this chemical causes the chicken to lay eggs.

People are learning a lot about internal clocks. Farmers have even learned how to fool a chicken's internal clock so that the chicken lays more eggs!

QUESTIONS

1. A *signal* is a
    a. chemical.
    b. message.
    c. clock.

2. The clock described in the story is called internal because it
    a. is inside the plant or animal.
    b. never stops running.
    c. produces a signal.

3. In this story, one signal would be
    a. cold weather.
    b. a chemical.
    c. falling leaves.

4. In the chain of events below, what is missing?

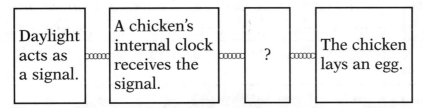

    a. The chicken is fooled by the farmer.
    b. The chicken makes a chemical.
    c. The chicken sees the daylight.

5. What do you think farmers do to make chickens lay more eggs?
    a. They turn on lights in the chicken house at night.
    b. They feed the chickens a special chemical.
    c. They keep the chickens in a cooler place.

The Big One

What is the largest animal ever to live on Earth?

If your answer was the great blue whale, you are right. The blue whale is an air-breathing mammal (măm'əl). A mammal belongs to the group of animals that have fur or hair on their bodies. The females of this group produce milk for their babies. A baby blue whale is almost 8 yards long when it is born, and it drinks about 100 gallons of milk a day from its mother. Baby whales are among the fastest-growing animals in the world.

A grown blue whale is about 100 feet long and about 40 times heavier than an elephant. Its body is covered with thick layers of blubber, or fat, and it spends most of its life in the freezing cold waters of the Antarctic Ocean. Blue whales travel in groups called *pods* (pŏdz) and eat plankton (plăngk'tən), tiny plants and animals that float near the surface of the ocean.

The blue whale has been hunted for its meat and blubber, almost to the point of *extinction* (ĭk stĭngk'shən). Now, people are trying to save this giant of the seas.

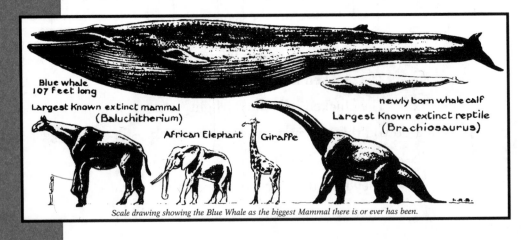

Blue whale 107 feet long

Largest known extinct mammal (Baluchitherium)

African Elephant

Giraffe

newly born whale calf

Largest known extinct reptile (Brachiosaurus)

Scale drawing showing the Blue Whale as the biggest Mammal there is or ever has been.

1. Tiny plants and animals that float near the surface of the ocean are called

   a.   pods.

   b.   blubber.

   c.   plankton.

2. Which of the following statements is *true*?

   a.   Blue whales travel in groups called pods.

   b.   Young blue whales do not grow very fast.

   c.   An elephant weighs more than a full-grown blue whale.

3. A baby whale lives *mostly* on _____ during the first year and a half of its life.

   a.   its mother's milk

   b.   tiny ocean plants

   c.   small sea animals

4. What protects the blue whale from the freezing temperatures of the Antarctic Ocean?

   a.   its great length of 100 feet

   b.   the plankton that it eats

   c.   its thick layers of blubber

5. Under which of the following headings would you list the blue whale?

   a.   Mammals of the Ocean

   b.   Extinct Animals of the Ocean

   c.   Cold-Blooded Animals of the Antarctic

Sharks!

The great white shark is an ancient and deadly fish.

A *shark* (shärk) is a meat-eating fish that is found all over the world, especially in warm waters. Scientists believe that sharks have lived in the world's oceans for millions of years. A shark has no bones. Its skeleton is made up of *cartilage* (kär′ tl ĭj), a tough material that stretches like an elastic band.

There are many kinds of sharks. The picture below shows a great white shark. Its body is shaped like a torpedo, and it can grow to be more than 40 feet long. It has two rows of sharp, pointed teeth. The great white shark has been called the perfect fish. It is a great swimmer. It eats large fish with a single gulp and has no natural enemies. The shark also preys on seals. Sometimes it mistakes a person for a seal. But the shark does not like the taste of people. It usually spits them out. But that first bite can do a lot of damage to the person.

1. What is *cartilage*?

   a. the bone in a shark's skeleton

   (b.) a tough, elastic material

   c. a kind of shark

2. Its torpedo shape helps the white shark to

   a. stretch like an elastic band.

   (b.) swim with great speed.

   c. blow up fishing boats.

3. How are all sharks alike?

   a. They are white.

   (b). They eat meat.

   c. They have bony skeletons.

4. Fishers may well fear the white shark's

   a. huge tail.

   b. pointed skeleton.

   (c). sharp teeth.

5. The white shark is a perfect fish because

   a. it can swim.

   (b). it does everything well.

   c. it has teeth.

The Flashlight Fish

Have you ever seen a fish with its very own lights?

Some fish have built-in lights. That is, the lights are actually a part of the fish's body! One such fish lives in the Red Sea. It is called the "flashlight fish" because it has special body parts, or *organs* (ôr′ gənz), that give off light.

The flashlight fish has one light organ underneath each eye. These lights are bright and greenish in color, and they are always "on." But a flashlight fish can conceal, or hide, its lights by raising an extra piece of skin up over each organ. Then, when the fish lowers the skin, the lights can be seen again.

What purpose do these special lights have? The light organs help the fish locate food at night along the Red Sea reefs. They also help the fish escape from its enemies. Usually, the lights are flashed "on" and "off" about once every 20 seconds. But when the fish is upset, it will blink its lights about 75 times in one minute. Can you imagine 20 or 30 flashlight fish blinking their lights all at the same time?

QUESTIONS

1. In this story, the word *organ* means
    a. an extra piece of skin.
    b. a special body part.
    c. a kind of eye.

2. The flashlight fish uses its lights to hunt for food and to
    a. change color.
    b. conceal its eyes.
    c. escape from enemies.

3. The flashlight fish gets its name from its
    a. greenish color.
    b. special organs.
    c. unusual eyes.

4. When a flashlight fish is upset, its lights will
    a. be covered.
    b. flash faster.
    c. turn bright green

5. If a flashlight fish were looking for food at night it would *probably*
    a. lower its special piece of skin.
    b. turn green in color.
    c. look for flashing lights.

The World's Largest Rodent

What kind of animal looks like a giant rat, swims like a duck, and eats water plants and vegetables?

It is a *capybara* (kăp' ə bä' rə), the world's largest rodent. A full-grown capybara can be up to 4 feet in length and weigh more than 100 pounds! This strange-looking rodent has a large, flat head and small, round ears.

Like all rodents, the capybara has strong front teeth that are wide and long and are used to gnaw its food. Its eyes are set so far back on its head that a capybara can't look at anything with both eyes at once.

The capybara's body is fat and covered with hair, and the rodent has almost no tail at all. It has short strong hind feet and even shorter front feet, which are webbed like a duck's.

Watching capybaras play is great fun, for they love to swim and roll together in warm water. They are friendly animals, but in the United States you will see them only in a zoo. Most capybaras make their homes in parts of South America where the water and the weather are often warm.

1. The *capybara* is the world's largest
    a. duck.
    b. rat.
     c. rodent.

2. The body of the capybara is covered with
     a. hair.
     b. fur.
    c. feathers.

3. There aren't many capybaras in the United States because our weather gets too
    a. rainy.
    b. cold.
    c. warm.

4. Which phrase best describes the shape of the capybara's body?
    a. long and thin
    b. round and fat
    c. small and flat

5. Why can this strange-looking rodent swim like a duck?
    a. It has a flat head.
    b. Its feet are webbed.
     c. It has almost no tail.

Sea Animal May Help Dentists

From the sea comes a new idea for dentists.

Whirr. Bzzz. Zzzt. At last, the dentist is finished drilling away the *decayed* (dĭ kād′), or bad, part of the tooth. Now there is a large hole, or *cavity* (kăv′ ĭ tē), that has to be filled. The filling the dentist uses should be made of very strong material. It will have to stay in the cavity no matter what kind of food the person eats. The filling will have to remain hard in a place that is always wet. Right now, the fillings dentists use sometimes fall out.

But at the seashore, far from the dentist's office, is an animal that may help solve the problem. This animal is the *mussel* (mŭs′ əl), a tiny sea animal that lives inside two shells. The mussel attaches, or fastens, itself to rocks in the water. It does this with a sticky material that it makes. This sticky material gets very hard and stays strong even in the water. Dentists are performing tests to find out what is in the mussel material and what makes it so strong.

1. According to the story, a *cavity* is a

    a.   strong material.

    b.   filling.

    c.   large hole.

2. A *mussel* is a kind of sea

    a.   shell.

    b.   rock.

    c.   animal.

3. Why does a mussel produce the special material?

    a.   to hold its shells together

    b.   to fasten itself onto rocks

    c.   to fill large holes

4. Why would the material from mussels make a good dental filling?

    a.   It would be easy to produce.

    b.   It always remains sticky.

    c.   It stays hard when wet.

5. Dentists are interested in the mussel material because it is

    a.   wet.

    b.   sticky.

    c.   strong.

Invite a Long Lifetime

Your heart is an *organ* (ôr′ gən), a part of your body that performs a very special job. Your heart's job is to pump blood through your body, and it must pump for a lifetime. The way you treat your heart now will affect whether or not it will continue to pump well or develop heart disease. Heart disease is caused by almost anything that weakens the heart or keeps it from doing its job properly. Both young people and old people are victims of this disease.

What can you do to avoid trouble and keep this important organ in top shape? For one thing, you can cut down on the amount of salt you eat. Salt can cause people to have high blood pressure. Also, don't eat too much butter or too many fatty foods. The excess fat will settle on the walls of the *arteries* (är′ tə rēz) that lead to the heart. This makes it difficult for blood to flow through them. And *do* take time each day to exercise. The proper amount of exercise makes the heart stronger.

1. The organ that does a special job is called your
   a. heart.
   b. arteries.
   c. body.

2. People with high blood pressure probably
   a. eat too many fatty foods.
   b. don't get enough exercise.
   c. eat lots of salty foods.

3. According to the story, heart disease can attack
   a. only older persons.
   b. persons of any age.
   c. only younger persons.

4. Your arteries cannot do their job if you
   a. eat too many fatty foods.
   b. don't get enough exercise.
   c. eat too many salty foods.

5. Keeping your heart healthy is important because it
   a. must pump for a lifetime.
   b. will keep your arteries strong.
   c. can help you avoid high blood pressure.

Protecting Your Health

A physical examination is a good way to check up on your health.

It is a good idea to have a *physical* (fĭz′ ĭ kəl) examination, or checkup, about once a year. The purpose of the checkup is to make sure that each part of your body is growing and doing its job properly.

During a complete physical examination, your body is checked from head to toe. The doctor uses different kinds of equipment, including a *stethoscope* (stĕth′ ə skōp′). This instrument is used to listen to the sounds made by your lungs and heart. A flat wooden stick called a tongue depressor is also used. With it, the doctor can check your tonsils and throat. The doctor checks different parts of your body for lumps or sore spots.

If you are having problems such as not sleeping or losing weight, this a good opportunity to discuss them with your doctor. It is also a good time to check on the shots you need to protect your body from illness.

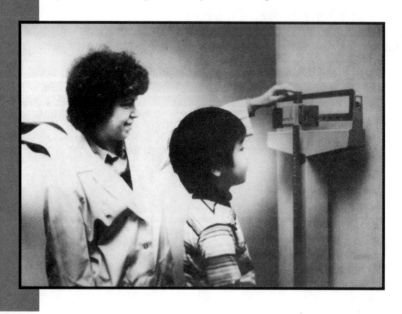

1. An instrument used to listen to the sounds made by your lungs and heart is called a _____.

2. The doctor may use a tongue depressor in order to _____.

3. You do not have to be sick to have a physical examination.

    a. True

    b. False

    c. The story does not say.

4. According to the story, during a physical examination what does the doctor check for?

    a. flat feet

    b. lumps or sore spots

    c. color-blindness

5. According to the article, shots

    a. hurt.

    b. will protect you from illness.

    c. are not necessary.

The Hardiest Weed of All

In spring, its sunny yellow head can be seen almost everywhere. This little yellow flower is a weed called the dandelion (dăn′ dlī′ ən). Once started, it spreads quickly and is very hard to control.

The dandelion is a perennial (pə rĕn′ ē əl), which means that its leaves and flowers die but its roots live on from year to year. Chopped off at the ground, this hardy perennial will grow again in one to two weeks and will often grow two plants in place of one.

The head of the dandelion is really a tight little cluster of 150 to 200 flowers. When the flower cluster dies it forms a fluffy gray ball, which holds the plant's seeds. There is one seed for every flower in a cluster. A light wind can carry the seeds about 50 feet. Is it any wonder that gardeners think of the dandelion as a pest?

1. A plant whose leaves and flowers die but whose roots live on from year to year is called a

    a.   seed.

    b.   perennial.

    c.   cluster.

2. When chopped off at the ground, the dandelion will

    a.   die in one to two weeks.

    b.   grow again in one to two weeks.

    c.   grow again the following spring.

3. The head of the dandelion is really a

    a.   cluster of yellow flowers.

    b.   fluffy gray ball.

    c.   group of small yellow leaves.

4. What makes the dandelion hard to control?

    a.   Its flowers die and come back the next year.

    b.   Its leaves can grow in one to two weeks.

    c.   Its seeds get scattered easily.

5. Under which of the following headings would you list the dandelion?

    a.   Gray Flowering Plants

    b.   Seedless Perennials

    c.   Perennial Pests

Red Seaweed

An ocean weed is used in making many products.

What do we get from the oceans of the world? All kinds of fish, yes, but we also get a plant called *seaweed* (sē'wēd'). One type of seaweed is red in color and grows to be about one yard in length.

When we think of a weed, we think of a useless, unwanted plant. But seaweed can be used for many purposes. One important product obtained from red seaweed is *agar* (ā'gär). Scientists use agar in many ways, including the making of certain foods such as candy. Agar is also used in making some medicines.

One kind of red seaweed that is rich in agar is found off the island of Japan. Divers collect this seaweed and load it onto boats. When a boat is full it returns to shore, where the seaweed is unloaded and spread out to dry in the sun. When completely dry, it is gathered up into bundles and shipped to a factory. At the factory, agar is made from the seaweed.

The next time you buy candy, read the ingredients on the wrapper. You may find agar listed!

1. What is *agar*?

    a.   red seaweed

    b.   a product made from seaweed

    c.   a kind of medicine

2. The red seaweed used to make agar is gathered by

    a.   fishers.

    b.   scientists.

    c.   divers.

3. Besides growing in the ocean, seaweed is different from most weeds because it can

    a.   grow tall.

    b.   be useful.

    c.   turn red.

4. Before we can get agar from it, the seaweed must be

    a.   dyed red.

    b.   completely dried.

    c.   sent to Japan.

5. What is the *main idea* of this story?

    a.   Using seaweed is too much trouble.

    b.   Seaweed can be very valuable.

    c.   Japan depends on seaweed.

Life in the Forest

There are many kinds of animals that live in the forest, such as mice, rabbits, squirrels, owls, and hawks. Animals in the forest depend upon one another for food.

Many small forest animals, such as mice, eat green plants for food, and then they, themselves, are eaten by larger animals. This is called a food chain.

A food chain begins with the sun. The green plants in the forest use the sun's energy to make food for themselves. This process, or action, is called *photosynthesis* (fō' tō sĭn' thĭ sĭs)

Forest plants produce flowers and seeds that are eaten by mice and other animals. The mice get their energy from the plants, and the mice become food for a larger animal, such as the hunting hawk. So the energy from the sun passes from the plants to the mice to the hawk. A break in any link of this food chain means that some animal may not have enough food to survive.

QUESTIONS

1. *Photosynthesis* takes place when

    a.   green plants use the sun's energy to make food.

    b.   forest animals eat the food made by green plants.

    c.   the food chain is broken.

2. What is the *first* link in a food chain?

    a.   the sun

    b.   green plants

    c.   forest animals

3. Animals sometimes cannot find enough food because

    a.   there is very little food in a forest.

    b.   photosynthesis stops.

    c.   there is a break in the food chain.

4. What does the food chain show about living things?

    a.   that they depend upon each other

    b.   that small animals eat as much as large animals

    c.   that plants with flowers are most important

5. Which of the living things below fills the missing link in this food chain?

    a.   mice

    b.   green plants

    c.   owls

Tree Rings

You can find out many things about a tree by studying its rings.

Each year, from top to bottom, a tree wraps a new layer of wood around itself. If you cut across a tree trunk or branch, you would see that these layers of wood look like rings. By counting *tree rings* (trē rĭngz), you can tell how old the tree is. If you count ten rings, for example, the tree is ten years old.

Looking at the picture shown below, you can see that each tree ring has a dark section and a light section. The light wood grows in the spring, the dark wood in the summer. Observe also that some rings in the picture are wide and others are narrow. Trees grow wide rings during the years when they get a lot of moisture and sunlight. By examining tree rings, you can tell what the weather may have been like many years ago. Giant sequoia trees, such as the one in the photo below, grow in northwestern California.

Some of these sequoias have more than 3,000 tree rings. Can you guess how old such trees would be?

1. According to the story, a *tree ring* is a

    a. branch.

    b. layer of wood.

    c. giant sequoia.

2. A tree ring is added to a tree

    a. every ten years.

    b. every year.

    c. every 3,000 years.

3. The light part of a tree ring

    a. gets dark in the winter.

    b. tells the tree's age.

    c. grows in the spring.

4. From the story, you can tell that

    a. all tree rings look the same.

    b. most trees do not grow tree rings.

    c. tree rings add to the tree's width.

5. Look at the drawing below. If this were a slice from a tree trunk, how old
    would the tree be?

    a. five years old

    b. five months old

    c. six years old

Watch Out for Poisonous Plants

Don't touch! That plant is poisonous!

Did you know that in the United States there are almost 400 kinds of *poisonous* (poi′ zə nəs) plants? A poison is any substance, or material, that causes injury, illness, or death. Perhaps you have heard of poison ivy and poison oak. These plants are found in North America. By touching them, a person can get an annoying and sometimes painful skin condition that is called dermatitis (dûr′mətī′tĭs). Dermatitis causes the skin to get red and itchy and develop watery blisters.

Poison ivy grows as a bush or a vine. It has small green flowers and white berries. The plant's leaves grow in groups of three, and they can be smooth and shiny or hairy. What should you do about poisonous plants? Find out what kinds of plants grow in your neighborhood. Learn to recognize plants such as poison ivy and poison oak, which grow freely in many places.

1. A *poison* is any substance that

    a. causes illness, injury, or death.

    b. is found in plants with groups of three leaves.

    c. grows freely on vines in North America.

2. One way to recognize poison ivy is by its

    a. hairy flowers.

    b. white berries.

    c. four leaves.

3. When people get dermatitis from poison ivy, what might they notice *first*?

    a. smooth, shiny skin

    b. watery blisters

    c. red, itchy skin

4. The *main purpose* of this story is to tell the reader how

    a. to find poisonous plants.

    b. important it is to recognize poisonous plants.

    c. to treat diseases caused by poisonous plants.

5. Under which of the following headings would you list poison ivy?

    a. A Leafless Vine

    b. A North American Plant

    c. A Poisonous Fruit

The Misunderstood Spider

How much do you really know about spiders?

Spiders are misunderstood animals. For one thing, they are not really insects, as most people think. Insects have six legs, while spiders have eight. And spiders have only two body parts, the head and the abdomen (ăb′ də mən). Insects have a third part, called the *thorax* (thôr′ăks′).

Of the 30,000 different kinds, or *species* (spē′ shēz′), of spiders, only some spin webs. All spiders produce silk, but each species has its own way of using the silk. The web spinners weave silky webs to catch food. The best-known web is the circular, or round, web spun by the garden spider. Other spiders may make nets to drop over their victims. One species even sends out a single sticky strand of silk like a fishing line to catch its food. When a fly gets stuck to the end of the line, the spider hauls in its "fish."

1. Each different kind of spider is a different _____.

2. According to the story, web spinners use their silky webs to

    a. sleep in.

    b. catch food.

    c. hide themselves.

3. How many body parts do spiders have?

    a. two

    b. three

    c. six

4. The story leads you to believe that some spiders eat

    a. fish.

    b. flies.

    c. people.

5. Another title for "The Misunderstood Spider" might be

    a. "Not Really an Insect."

    b. "A Useless Insect."

    c. "Common Garden Insects."

EARTH-SPACE SCIENCE

Have you ever heard of an underwater well? There are many such oil wells. Oil forms underground. It may be found under farmland, jungles, swamps, or mountains. Large amounts of oil are found under the bottom of the ocean. This oil well in the Gulf of Mexico has created a gusher. Gushers are rare today, because scientists have found ways to control the flow of oil as it comes out of the well.

WORDS TO KNOW

Tornado Warning!
occurs, happens
funnel, a slender tube larger at the top than to the bottom

Nature's Deep Freeze
wooly mammoth, an extinct elephant with hairy skin and long tusks curving upward
intact, nothing missing, left whole

Fossils of Ocean Animals
marine, found in the sea
fossils, hardened remains of plants or animal life

Warm-Blooded Dinosaurs?
paleontologists, scientists who deal with prehistoric life through the study of fossils
internal, inside of
fossilized, turned into a fossil (See "Fossils of Ocean Animals.")

Jupiter's Moons and How They Travel
planet, a heavenly body that circles the sun
solar, having to do with the sun (our sun and the sun's planets)
sequence, the order in which some thing happens
counterclockwise, turns or spins in a direction opposite the way the hands of a clock turn

The Moon Named Io
astronomer, scientists who study the heavens
patches, different parts of a surface

The Planet Mercury
craters, large holes or depressions
NASA, The National Aeronautical and Space Administration, an agency of the U.S. government.

Fog
droplets, tiny drops
blanket, covering

Tornado Warning!

There can be danger in a dark spring sky.

People stop what they are doing and quickly go to a shelter, or safe place, when they hear the siren. The siren is a warning that a *tornado* (tôr nā′ dō) is coming.

A tornado is a storm that usually occurs in the spring and early summer months in certain parts of the world. A tornado covers only a small area. But it can destroy everything in its narrow path because the winds of a tornado spin very fast.

What causes the winds to move so fast? When cold air sinks and when wet air rises, winds begin to spin. As the winds spin faster, more warm air is pulled upward. Then cold air moves in under the warm air. If this action continues over and over, a large, funnel-shaped cloud soon rises up in the sky. The pointed end of the funnel rushes along the ground, sucking up everything in its way. A tornado can destroy an entire town and kill many people. But people who hear and obey the warning can be safe.

1. A *tornado* can easily be identified by its

    a.   slowly rising temperatures.

    b.   large, funnel-shaped cloud.

    c.   noisy, siren-like winds.

2. According to the story, a tornado usually takes place in the _____ or early summer.

3. According to the story, which of the following statements about a tornado is true?

    a.   It always covers a very wide area.

    b.   It can destroy everything in its narrow path.

    c.   It usually occurs at any time all over the world.

4. Paragraph 3 leads you to believe that

    a.   warm air is lighter than cold air.

    b.   warm air is pulled downward by spinning winds.

    c.   when cold air moves up, warm air comes in.

5. What part of a tornado actually causes the damage?

    a.   the cloud

    b.   the spinning winds

    c.   the warm, wet air

Nature's Deep Freeze

What happened during the Great Ice Age?

Thousands of years ago, the winters began to get colder, and the summers grew shorter until there were no summers at all. This was a period known as the Great Ice Age. Snow piled up 15 stories high over much of the Northern Hemisphere. Under the weight of so much snow, an ice pack formed and began to push forward. At that point, *a glacier* (glā′ shər) was born.

As it moved slowly down the mountains, the glacier buried unlucky animals that were trapped in cracks in the earth. There they remained, perfectly frozen, for thousands of years. Then, slowly, the climate changed and the glacier began to melt. About 100 years ago, scientists made an amazing discovery. They found a woolly mammoth that had been preserved intact in the deep freeze for 10,000 years!

1. A mass of thick moving ice is called a _____.

2. What happened during the period in history called the Great Ice Age?

    a. Winters began to get colder.

    b. The climate began to get warmer.

    c. A huge mountain of snow began to melt.

3. According to the story, which of the following statements is true?

    a. Animals buried by the glacier remained frozen for thousands of years.

    b. It takes very little snow to form an ice pack.

    c. A glacier moves very quickly as it pushes forward.

4. Because our weather today is warmer than it was thousands of years ago,

    a. there are many more woolly mammoths.

    b. there are fewer glaciers.

    c. it no longer snows in the Northern Hemisphere.

5. The story says that the woolly mammoth had been frozen for 10,000 years. Therefore,

    a. scientists must have discovered the animal 10,000 years ago.

    b. the animal must have been 10,000 years old when it was caught in the ice.

    c. the glacier had to be at least 10,000 years old.

Fossils of Ocean Animals

Why do scientists think the middle of North America was once under a warm sea?

Today, Missouri, Kansas, Nebraska, Iowa, and Indiana are hundreds of miles from an ocean. Yet scientists have found evidence, or proof, that marine (mə rēn'), or sea, life once lived in each of these five states. The evidence is in the form of *fossils* (fŏs'əlz), the remains of plants or animals that lived millions of years ago and are often found preserved in certain kinds of rock.

Fossil hunters discovered the remains of marine animals without backbones, including sea snails and clams. Some fossils were perhaps 500 million years old! They also found fossils of marine life with backbones, such as sea turtles, crocodiles, and bony fish.

Some fossil marine animals, such as starfish and clams, are easy to recognize. But others look like beautiful flowers with stems and buds. These sea lilies, or *crinoids* (krī'noidz), still live today in warm sea waters and use their petal-like arms to catch tiny sea animals for food. The state of Iowa is famous for its fossil crinoids.

1. *Marine* life is life that exists _____.

2. Which of the following is proof that the middle part of North America was once underwater?

   a. marine fossils

   b. certain kinds of rock

   c. live crinoids

3. Crinoids are known as sea lilies because they are

   a. actually underwater flowers.

   b. so easy to recognize.

   c. shaped like flowers.

4. The fossil crinoids found in Iowa show that long ago the state was

   a. probably under a sea of warm water.

   b. covered with layers of rock.

   c. only a few hundred miles from the ocean.

5. Under which of the following headings would clams belong?

   a. Marine Life without Backbones

   b. Underwater Plant Life

   c. Fossils with Stems

Warm-Blooded Dinosaurs?

We all have seen movies showing dinosaurs as big and slow-moving. This is because many *paleontologists*, the scientists who study dinosaurs, were sure that animals that big had to be cold-blooded.

What is a cold-blooded animal? It is a reptile, like a snake or lizard, with a simple heart and no internal way to warm itself. It has to sit in the sun for hours before it can warm up enough to move about. If dinosaurs were cold-blooded, they would have moved slowly. but there were fossil footprints of fast-moving dinosaurs!

Cold-blooded or warm-blooded? No one had an answer to the question until early in 2000. Digging in South Dakota, scientists found the fossil of a *Thescelosaurus*, a 66-million-year-old plant-eating dinosaur. Usually only hard material, such as bone, can become fossilized. This time, though, the scientists also found the dinosaur's fossilized heart. It wasn't a reptile's heart at all. Instead, it had four chambers, like those of mammals, such as dogs, cats, and human beings!

The question may finally have an answer. The dinosaurs, at least some of them, seem to have been warm-blooded.

1·07 × 2 m5

1. Scientists who study dinosaurs are called _____.

2. What are some cold-blooded animals?

    a.   dogs and cats

    b.   birds

    c.   snakes or lizards

3. According to the story, if dinosaurs were cold-blooded they would move _____.

4. Before the year 2000, scientists thought that some dinosaurs might be warm-blooded because

    a.   they found dinosaur blood.

    b.   they found fossil footprints of fast-moving dinosaurs.

    c.   they found a dinosaur heart with four chambers.

5. If scientists find more fossilized dinosaur hearts with four chambers, it will prove

    a.   that all dinosaurs were warm-blooded.

    b.   that all dinosaurs were fast moving.

    c.   that at least some dinosaurs were warm-blooded.

Jupiter's Moons and How They Travel

The many moons of Jupiter travel around the planet in different directions.

Jupiter is the largest planet in our solar system. Over the years, scientists have found that Jupiter has its own small solar system. Earth has one moon. Jupiter has at least seventeen and probably more.

Since there are so many moons, scientists began to number them. The numerals tell the sequence, or order, in which the moons were found. The scientists were slower to name the moons. Now, however, all of Jupiter's moons have a name as well as a number.

The first five moons to be discovered are known as the "inner moons." But they are not the closest to the planet. The closest is only 76,560 miles away from Jupiter. All the inner moons circle the planet in a counterclockwise (koun'tər klŏk 'wīz') direction, that is, opposite of the hands of a clock.

Jupiter's middle group of moons is at least 6,660,000 miles from the planet. They also move in a counterclockwise motion. The four farthest moons are at least 12,420,000 miles away. These are called "outer moons." They circle in a clockwise motion.

How many more moons do you think will be discovered?

1. Things that travel in the same direction as the hands of a clock are said
    to be traveling in a
    a. clockwise direction.
    b. counterclockwise direction.
    c. different direction.

2. Jupiter's _____ group of moons travel in a clockwise direction.
    a. inner
    b. middle
    c. outer

3. The numbers given to Jupiter's moons tell
    a. the order in which they were discovered.
    b. the order in which they travel.
    c. the order of their distance from Jupiter.

4. According to the story, which of the following statements is true?
    a. None of Jupiter's moons have names.
    b. Most of Jupiter's moons circle clockwise.
    c. Jupiter's inner moons were discovered first.

5. Use the three lines below to record how far away each group of moons
    is from Jupiter.
    Inner Group: _____ miles
    Middle Group: _____ miles
    Outer Group: _____ miles

The Moon Named Io

Galileo spacecraft takes pictures of Jupiter's moons.

In 1996 the spacecraft *Galileo* took thousands of pictures of the planet Jupiter. It also took many pictures of Jupiter's moons. A moon is a *satellite* (săt′l ĭt′), that is, a body that travels in a path around a larger body, or planet.

One of Jupiter's moons, or satellites, is named Io (ī ō) and is about the same size as Earth's moon. Galileo, an astronomer, discovered the moon in 1610. Unlike our moon, Io has almost no craters, or large holes, on it.

Scientists once thought that Io would be like Earth's moon—cold and without much color. But photographs sent back by *Galileo* show that Io is a bright orange-red color with patches of white here and there. The photos of Io also show that there are many volcanoes on it and that quite a few of these are active. That is, gases and other materials shoot out of the volcanoes.

The volcanoes on Io appear to be more powerful than those found on Earth. So far, Jupiter's Io and our planet Earth are the only known bodies in space that have active volcanoes.

1. A body that travels in a path around a larger body, or planet, is called a

    a.   volcano.

    b.   satellite.

    c.   crater.

2. Earth's moon is about the same size as

    a.   Jupiter.

    b.   Io.

    c.   *Galileo*.

3. Scientists once believed that

    a.   Earth was one of Jupiter's moons.

    b.   Io was cold and had little color.

    c.   Earth's moon had active volcanoes on it.

4. Besides being bigger, how is Jupiter *unlike* Earth?

    a.   It has more than one moon.

    b.   It is red with white patches.

    c.   It has active volcanoes.

5. What is it about Earth and Io that makes them different from any other bodies in space?

    a.   They both have active volcanoes.

    b.   They both have many craters.

    c.   They are both bright red in color.

The Planet Mercury

What is the smallest planet in our solar system?

Mercury is the closest planet to our sun. It is one of the hottest planets in our solar system. During the day it is over 800 °F. This is hot enough to melt lead! At night it drops to below -300 °F.

A day on Mercury lasts about 1,416 Earth hours. Mercury is the fastest planet in our solar system. It takes just 88 days to revolve around the sun. Earth takes about 365 1/4 days to make the trip.

Mercury is hard to see because it is so near the sun. Not very much was known about it until 1974. That was

when a spacecraft was sent to take pictures of it. The pictures showed Mercury has mountains, hills, and valleys. It is covered with rocks. There are many *craters* (krā′tərz), too. Craters are large holes made when objects from space crash into planets.

In the year 2004, NASA will send another spacecraft to Mercury. NASA is in charge of U.S. space exploration. The new spacecraft will be named *Messenger*.

1. Craters are

    a.   tunnels in the Earth's crust.

    b.   large holes made when space objects crash into planets.

    c.   boxes used to ship spacecraft.

2. How many days does it take for Mercury to revolve around the sun?

    a.   88

    b.   -300 °F

    c.   over 1,000

3. How long is a day on Mercury in Earth hours?

    a.   10 of our days

    b.   over 1,000 hours

    c.   same as on Earth

4. How long is a year on Mercury?

    a.   same as on Earth

    b.   1,416 days

    c.   88 days

5. What would be a problem with humans trying to live on Mercury?

    a.   It travels so fast, people would fly off.

    b.   The year would be too short.

    c.   It is so hot they would burn up.

Meteorites: Space Rocks That Fell to Earth

What are meteorites? How can you tell them from Earth rocks?

A *meteorite* (mē′tē ə rīt′) is rock-like matter that has fallen to Earth from space. Most meteorites look just like Earth rocks, but a meteorite the size of an orange weighs much more than an Earth rock of the same size. Meteorites can be many shapes and sizes. They can be as small as a peanut or as large as a truck.

Most meteorites are dark brown on the outside, and since most of them contain metal, they are silver or gray inside. Meteorites found on Earth are almost all metal inside and are known as iron meteorites. Other meteorites are made of stone or are a mixture of stone and metal. Iron meteorites are found more often because they are the least likely to break apart while falling through space or when hitting Earth.

Most stony Earth rocks do not contain metal. So if you think you have found a meteorite, check it with a magnet. If the magnet sticks to the rock, you may have found a space rock that has fallen to Earth!

Nickel-Iron ... rite (Siderite)
SANDIA MTS., New Mexico
Found about 1915
Total known weight 14 kg. (31 lb.)

1. Rock-like matter that has fallen to Earth from space is called a

   a. magnet.

   b. meteorite.

   c. metal.

2. A meteorite the size of an orange will weigh _____ Earth rock of the same size.

   a. less than

   b. much more than

   c. about the same as

3. If you found a meteorite, what color would it probably be?

   a. silver

   b. gray

   c. brown

4. A magnet will help you tell a stony Earth rock from a meteorite, because the magnet will

   a. stick to the stony Earth rock.

   b. stick to the meteorite.

   c. show which rock weighs more.

5. The biggest meteorites found are likely to contain more

   a. iron than stone.

   b. silver than iron.

   c. stone than metal.

Fog

We usually think of *fog* (fôg), with its cloud-like, moist air, as scary and troublesome. But fog can be good as well as bad. The fine water droplets in a fog layer gently moisten plants and soil.

Winter fog forms as a result of warm raindrops falling through colder air. Winter fog sometimes acts as a blanket that keeps the warm air "tucked" close to the ground. Fog may protect gardens from frost and keep buildings warm. Then less heating fuel is needed. In summer, a fog may block out sunshine, keeping the air at the ground level cool and moist.

Unfortunately, fog can also be dangerous. Heavy fog makes it difficult to see more than a few feet. Also, fog changes the way that sound travels. It causes sound waves to bounce up so high in the air that they skip over large areas without being heard. This problem is very noticeable

over large bodies of water. Sometimes foghorns and breaking waves cannot be heard. Danger can be close at hand, but neither sight nor sound gives a warning.

1. *Fog* is the word used to describe a cloud-like layer of air made up of fine
   _____ droplets.

2. Winter fog is formed when
   a. warm raindrops fall through colder air.
   b. cold air sinks to ground level.
   c. warm air acts as a blanket.

3. A heavy fog makes it difficult to
   a. keep the ground moist.
   b. see more than a few feet.
   c. keep warm air at ground level in winter.

4. Which of the following describes one way in which fog is good?
   a. Fog acts as a watering system for plants and soil.
   b. Fog decreases the noise level over large bodies of water.
   c. Fog develops only during the summer months, when the air is warm.

5. Which one of the following people would probably welcome fog?
   a. a sailor
   b. a gardener
   c. a driver

PHYSICAL SCIENCE

Windmills were once used to pump water on many American farms. When electricity became available, farmers replaced their windmills with electric pumps. Today, the rising cost of electric power is causing people to think again about using the power of the wind. New windmills are being built. Some scientists believe that the wind will once more be an important source of power in places where the wind blows steadily and often.

WORDS TO KNOW

Using Science to Save Our Banner

anthem, a song

Smithsonian Institution, the national history museum in Washington, D.C.

grime, dirt

Robots See, Touch, and Do

program, a plan of actions to be performed by a computer

weld, to join pieces of metal by heating

supervise, to manage, direct

pace, the speed of movement in a set time

Time Machines

gravity, the force that attracts objects to one another

warp, to bend

mass, bulk, size

On the Lunch Trail

crustacean, shrimps, crabs, barnacles, and lobsters

antennae, the pair of moveable sense organs that stick out of the head

plume, a large feather or a group of feathers. In the story, the odor trails that are shaped something like feathers.

robot, a mechanical device that works by remote control

Eating Metal

gnaw, to bite and wear away bit by bit

rivets, metal bolts inserted through holes to hold pieces of metal together

Gravity

force, something that causes, changes, or stops the motion of an object

acceleration, increasing speed

The Soapy Sandwich

combine, to join or mix together

evaporate, to disappear

mixture, two or more substances mixed together but not united chemically

Travel by Air

engineer, scientist who figures out how to make machines work

design, the drawing and planning of new things

takeoff, act of leaving the ground

detect, to discover or notice

Using Science to Save Our Banner

What does science have to do with a famous flag?

The flag that inspired Francis Scott Key to write "The Star Spangled Banner" is now 185 years old. The flag flew over Fort McHenry in Baltimore harbor during the War of 1812.

The huge flag was about half the size of a tennis court. It has hung in the Entrance Hall of the Smithsonian Institution in Washington for many years. There the flag was exposed to pollution, dirt, and oil floating up from machinery such as escalators. Now the red, white, and blue is marred with grime, rust spots, and small holes.

People at the museum have decided it's time to clean the flag. They want to find ways to preserve it so it won't fall apart. They have turned to science to find the best way to save the flag. The flag is made of wool. Testing has shown that light and oxygen weaken wool. A camera that can detect chemicals in space is being used to find out what kinds of stains are on the flag. Once they know what each stain is, the flag cleaners will know how to remove it. When the flag is clean, it will be displayed in a special case that protects it from light and oxygen.

1. "The Star Spangled Banner" is

    a.   our national anthem.

    b.   a British ship.

    c.   in Baltimore.

2. The flag was dirtied by

    a.   the war.

    b.   pollution.

    c.   escalators.

3. Wool fibers are weakened by

    a.   dirt.

    b.   light and oxygen.

    c.   rust.

4. In order to remove a stain,

    a.   you have to use soap.

    b.   you need oxygen.

    c.   you need to know what caused it.

5. Some of the small holes in the flag were probably caused by

    a.   light.

    b.   museum visitors.

    c.   oil.

Robots See, Touch, and Do

How many jobs can robots do?

Today there are thousands of robots at work. Robots are special machines that are controlled by a computer. The robot can be "taught," or programmed, by the computer to do a number of jobs. The computer controls the actions of the robot.

Robots do a number of jobs that are too boring or dangerous for people to do. Robots also do not get tired. They can produce work for a long time at a steady pace without making a mistake.

Most robots today are used in the car industry. One kind of robot is used to spray paint car and truck bodies. Another is used to weld metal door hinges.

How is a robot different from other machines? Robots are different from other machines because they are *versatile*. Other machines can usually do only one kind of job. Robots can be programmed to do more than one kind

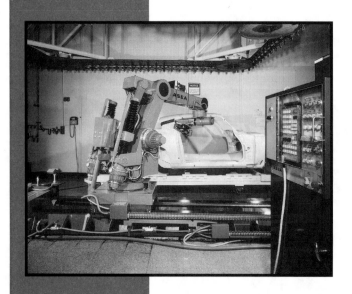

of job at the same time. As an example, the *Mars Polar Lander* spacecraft is a robot that was programmed to do several jobs. It was programmed to record sounds, dig up Martian soil, and take pictures of weather and atmospheric dust.

The use of robots will continue to grow. But it will be up to people to program and supervise the robots.

1. According to the story, robots can do

    a. only one kind of job.

    b. many different jobs.

    c. only jobs in the automobile industry.

2. What is used to program the action of a robot?

    a. a television set

    b. a computer

    c. a tape recorder

3. What kinds of jobs do robots do in the automobile industry?

    a. drive cars

    b. wash and wax cars

    c. spray paint cars

4. What does the word *versatile* mean in the fourth paragraph?

    a. to do many things

    b. to do nothing

    c. to do only one thing

5. Why do you think a robot might be used to fight a fire?

    a. The fire is only outdoors.

    b. The fire is too dangerous for firefighters.

    c. The fire is too small for firefighters.

Time Machines

No one has ever built a time machine. But the laws of physics do not rule it out. We usually think time goes forward. Physicists think it can run backward, too.

Black holes are dense, heavy points in space. Black holes have so much gravity, they warp space-time. The warping can cause *wormholes*. Wormholes are tunnels that link black holes. Scientists think it might be possible to go through a wormhole and come out in a different time.

Astronomer Frank Tipler has another idea for time travel. First a long, thin tube made of material ten times the mass of our sun would need to be built. The tube would need to be spun billions of times per minute. By traveling around the spinning tube, a traveler could end up billions of years from where the trip began.

Other scientists think cosmic strings could be used to make a time machine. Cosmic strings are thought to be very thin lines that twist space and time. No one has ever found one. But scientists think they might exist. By firing two long cosmic strings past each other at high speeds, they think a ship could travel around them in a figure eight. The ship could then end up anywhere in time.

1. Black holes are

    a.   dark tunnels in the Earth's crust.

    b.   heavy, dense points in space.

    c.   the same as wormholes, but they are very dark.

2. It is a fact that

    a.   black holes warp space-time.

    b.   cosmic strings can be used to climb through time.

    c.   wormholes have been used to travel in space.

3. Physicists think

    a.   a time machine will be made in the next decade.

    b.   time runs forward and backward.

    c.   building a tube ten times the mass of the sun is the best way to try time travel.

4. Based on the story, which do you think makes sense?

    a.   There might be a black hole between the moon and Earth.

    b.   You could get photographs of cosmic strings from the Internet.

    c.   If a spaceship came near a black hole, it would probably be sucked in by the gravity.

5. Based on the story, which do you think is logical?

    a.   It would be possible to build a time machine in your basement.

    b.   Scientists know about wormholes because astronauts had seen them from the moon's surface.

    c.   It would take enormous energy to travel through time.

On the Lunch Trail

How does lobster find lunch?

A lobster's search for lunch has helped scientists trail chemicals underwater. Scientists know that some lobsters weigh more than 20 pounds. How can a *crustacean* living on the ocean bottom eat so well? The light is dim. The water may be cloudy. The lobster's eyes are small.

The secret is in the lobster's *antennae*. A lobster uses its antennae to sniff out food. Scientists think a lobster compares signals received by the left and right antennae. The lobster moves toward the stronger odor. It continues to compare smells. The lobster follows the odor cloud or *plume* to lunch.

Trailing odor plumes may be easy for lobsters but not humans. Odors or chemicals do not spread evenly. Sometimes there are large odor plumes and sometimes only small patches. The water swirls odor away from a straight path.

Scientists need a computer to help trace underwater chemicals. So they are developing an underwater robot. *Robolobster* will move across the ocean floor searching for chemicals. The robot will sniff odor plumes and compare them. It will follow the strongest trail. Robolobster can already find a trail of saltwater in fresh water. Someday it may find other chemicals, even dangerous ones.

1. According to the story, a lobster is part of a group of animals called

    a. fish.

    b. crustaceans.

    c. seafood.

2. Lobsters search for food using their

    a. claws.

    b. eyes.

    c. antennae.

3. *Odor plumes* is another way of saying

    a. clouds of chemicals.

    b. dangerous chemicals.

    c. chemical feathers.

4. Which is not a reason underwater chemicals are hard to trace?

    a. Chemicals sink fast in cloudy water.

    b. The trails often break up into patches.

    c. Flowing water moves chemicals around.

5. Which problem might Robolobster help solve?

    a. finding polluted water

    b. finding out what lobsters eat

    c. learning which chemicals are dangerous

Eating Metal

What has no teeth but eats metal?

Metal objects last a long time, but most metals do wear out. Metal that is wearing out is *corroding*. The word *corrode* comes from the word *rodent,* meaning to eat or gnaw. Rats are rodents and they certainly can gnaw. But the thing that eats, or corrodes, metal is not a rat. The oxygen in air and water corrodes many metals, including iron and aluminum.

There is iron in steel. Both iron and steel turn red, or rusty, when they are corroded. If oxygen isn't kept out, these metals will rust away. So iron and steel objects are protected by special coats. The coats may be grease, paint, or other metals.

For years, car bumpers were coated with *chromium,* a shiny metal that corrodes slowly. The bumpers lasted longer than uncoated steel.

Corroding metals cause troubles in airplanes, too. When aluminum sheets are joined by rivets or overlapped, there are weak spots. Air and water eat tiny pits in the metal. But the pits are hard to see inside a plane. Repair crews worry they won't find the corrosion. So scientists have invented a clear paint that turns pink when corrosion starts. The repair crews hope this warning will help them spot trouble and slow oxygen attacks on airplanes.

1. The word meaning to *eat away or gnaw to pieces* is

    a. *coat.*

    b. *corrode.*

    c. *pit.*

2. How does paint prevent corrosion? It keeps

    a. metal shiny.

    b. metal smooth.

    c. oxygen away.

3. Paint which warns "corrosion is starting" will help

    a. scientists.

    b. car mechanics.

    c. airplane repair crews.

4. If you find a piece of rusty metal, it must be all or partly

    a. iron.

    b. aluminum.

    c. gold.

5. The *main idea* of this article is

    a. metals last forever.

    b. iron corrodes easily.

    c. many metals wear out.

Gravity

Which weighs more, a pound of lead or a pound of feathers? The answer depends on whether they are on the same planet.

Gravity is a force that causes objects to tend to move toward each other. Gravity is what makes objects have weight. Because of the pull of gravity, objects are attracted to the Earth. They do not float away from it. The gravity of all the planets in our solar system is what keeps them in their paths around the sun.

The effects of gravity are easy to see. But it is not clear exactly how gravity works. In the late 1600s, Sir Isaac Newton, an English scientist, observed an apple falling from a tree to the ground. This lead him to see gravity as a *pull* between Earth and an object. Newton showed that the strength of gravity is related to the size of the object. So, the planet Jupiter, which is much larger than Earth, has a stronger gravitational pull. That means that something that weighs one pound on Earth would weigh much more on Jupiter.

In 1915 Albert Einstein, a German-born scientist, added some new ideas to Newton's explanation of gravity. Einstein realized that gravity depended on other things beside size. It also depends on time, space, and acceleration. Later, when space travel became possible, many of Einstein's ideas about gravity were proved correct.

1. The force that attracts objects toward a planet is called

    a.   gravity.

    b.   orbit.

    c.   size.

2. Gravity is the reason objects

    a.   travel at certain speeds.

    b.   can go into outer space.

    c.   have weight.

3. According to the story, an object that weighs one pound on Earth would weigh

    a.   less than one pound on Jupiter.

    b.   more than one pound on Jupiter.

    c.   the same on Jupiter.

4. Scientists who have studied gravity

    a.   have always completely understood gravity.

    b.   have added to each others theories.

    c.   say we will never understand gravity.

5. According to the story, which happened last?

    a.   Space travel proved Einstein's ideas about gravity true.

    b.   Newton described gravity as a force that acted by pulling on objects.

    c.   Newton observed an apple fall to the ground.

The Soapy Sandwich

What kind of sandwich is fun to make, pretty to see, but soapy to taste? The answer is a soap bubble!

To make good soap bubbles, mix air with one part of water and two parts of liquid soap.

The water holds the two parts of liquid soap together. This is like the inside of a sandwich. The air around each soap bubble is like the bread around a sandwich. It holds the soap bubble together. This kind of soap bubble sandwich is what is called a *mixture*.

A mixture is anything that can be combined or mixed and then be taken apart again. How can you prove your soap bubble sandwich is a real mixture? Just boil it in a pot. The water will evaporate, or disappear, and leave the soap in the bottom of the pot!

If you add salt to water, you will have made another mixture. Do you want to prove it? Just boil the salty water in a pot. The water will evaporate and leave the salt in the bottom of the pot.

There are many other mixtures that you can make. But not many of them are as much fun as making a soap bubble sandwich.

1. The part of a soap bubble that acts like the bread around a sandwich is the
   a. water.
   b. air.
   c. liquid soap.

2. The inside of the soap bubble sandwich is
   a. the liquid soap and the water.
   b. the air.
   c. the salt.

3. A mixture is anything that
   a. will evaporate.
   b. can be combined and then taken apart again.
   c. will hold a soap bubble together.

4. The word *evaporate* means to
   a. boil.
   b. mix.
   c. disappear.

5. If you boil salty water in a pot,
   a. the water will evaporate.
   b. the salt and the water will become a mixture.
   c. the soap bubble will evaporate.

Travel by Air

The sky's the limit.

Many improvements have been made in the science of *aeronautics*. Aeronautics is the design and building of aircraft. Scientists and engineers in the field have been working on ways to make planes fly farther, faster, higher, more quietly and safely. They are working on planes that make quicker and steeper takeoffs from spaces as small as a parking lot.

Today, scientists use computers to design airplanes. A design can be tested on the computer to see how the new plane will fly. If the computer model doesn't work, the plane will never be built. This saves a lot of time and money.

In 1947 researchers produced the first plane to fly faster than the speed of sound. These planes are called "supersonics." We now have planes that can fly higher than ever before, even into outer space. The U.S. Space Shuttle can fly in outer space, return, and land on Earth.

New computer systems are being developed that can "learn" to fly airplanes. These systems will help pilots if the plane gets into trouble. The computer will be able to detect and solve a problem faster than a human pilot could. What sort of plane might be invented next? Where could you go in it?

1. The design and construction of aircraft is called _____.

2. Planes that can takeoff from smaller spaces would need _____ runways.

    a. shorter

    b. longer

    c. wider

3. The U.S. Space Shuttle is mentioned in the story as an example of an airplane that flies _____.

    a. faster

    b. quieter

    c. higher

4. According to the story, what are scientists using computers for?

    a. to figure out how much it will cost to design an airplane

    b. to test whether new ideas for airplanes will work

    c. to pilot the Space Shuttle

5. After reading this story, you could say that

    a. helicopters fly more quietly than airplanes.

    b. people are always working on ways to improve air travel.

    c. there will always be a need for longer and longer runways.

ENVIRONMENTAL SCIENCE

In this picture you are looking up into sequoia trees, among the largest and oldest living things on earth. Millions of years ago these trees grew in large forests throughout much of the world. There were many different kinds. Today there are only two true kinds of sequoia trees remaining, the redwood and the giant sequoia. Both are chiefly found in California. Environmentalists are fighting to save these trees.

WORDS TO KNOW

The Peregrine Falcon
decline, to lessen
efforts, attempts

Chemical Wastes
dumped, carelessly thrown away on top of the ground
pollution, dirtied soil, water, or air

Glacier Warning
shrink, to get smaller
disappear, to go out of sight
gravity, the force that tends to draw all objects to the center of the Earth

Rhinos Need Protection
medicine, materials used in the treatment and prevention of disease
vegetation, the plants of an area or region
snout, the nose and jaw of an animal
percent, a hundredth part
decades, periods of ten years
extinction, the fact of no longer existing
ornaments, anything used for decoration

Aquaculture, Another Way to Raise Fish
aquatic, growing or living in water
ornament, a special colorful object to wear

aquarium, a tank for keeping live water animals or plants
industry, business
mangrove, tropical trees or shrubs that grow along tidal shores

The Peregrine Falcon

Why are peregrine falcons returning to the wild?

Sometimes birds, like the peregrine falcon, eat poisons that are really meant for insects. Once there were peregrines all over the world. But by the 1960s, there were many fewer birds. What caused the number of peregrines to decline?

The problem began when a *pesticide* (pĕs′ tĭ sīd′) called DDT was sprayed on crops. A pesticide is an insect killer, and the pesticide DDT was used to kill insects that ruined farmers' crops. The poisoned insects were eaten by small birds like quail. Then the peregrine falcons ate the quail.

This continued, and as the DDT filled the body of the falcon, something terrible happened. When the falcons laid their eggs, the shells of the eggs were very thin. So before the baby peregrines were ready to hatch, the shells broke and the baby peregrines died.

Major efforts were made to increase the number of peregrines. Laws against the use of DDT and other pesticides were passed in the United States. Scientists

raised peregrine falcons and sent them back into the wild. The birds found places to live and lay their eggs. The birds are increasing their numbers. Now there is a good chance that the peregrine falcon will be saved.

1. A *pesticide* is

    a. a baby peregrine falcon.

    b. an insect that eats crops.

    c. an insect killer sprayed on crops.

2. According to the story, farmers used DDT to

    a. protect their crops from insects.

    b. kill birds that were destroying their crops.

    c. destroy the quail.

3. Peregrine falcons were in danger of dying out because

    a. their young died when shells broke too early.

    b. they were being sprayed with DDT.

    c. quail were eating the baby peregrines.

4. The shells of the falcons' eggs were thin because

    a. insects were inside the shells.

    b. pesticides were sprayed on the shells.

    c. DDT from the mother's body went into the eggs.

5. Which happened *first*?

    a. The quail ate insects.

    b. The peregrines ate quail.

    c. The farmers sprayed crops.

Chemical Wastes

What can we do about dangerous chemical wastes?

Many of the things we use today were made because of the work of special scientists called *chemists* (kĕm'ĭsts). A chemist makes and uses special materials called *chemicals* (kĕm'ĭ kəlz), which are then used to make products such as paints, dyes, paper, and certain types of clothes. Often, there are some chemicals left over after these products are made. These leftover, or unused, chemicals are called *chemical wastes* (kĕm'ĭ kəl wāsts).

What happens to chemical wastes? For a long time, chemical wastes were buried in the ground or dumped. Chemical wastes are a major source of lake, river, stream, and ocean pollution. Certain chemical wastes can harm people and other living things.

People are working to pass laws to stop the dumping of chemical wastes. Some companies are looking for safer ways to dispose of them. Also, scientists are testing ways to use leftover chemicals to make other useful products. Some of the waste is put in drums and buried in safe places. Other waste is burned.

1. *Chemical wastes* are

    a.   materials for clothes.

    b.   leftover chemicals.

    c.   chemicals that have been used up.

2. According to the story, for a long time chemical wastes had been _____.

3. Today, we know that chemical wastes can

    a.   never be used again.

    b.   be harmful to all living things.

    c.   be dumped safely in rivers or streams.

4. In the story, one way to lessen the dangers from chemical wastes is

    a.   to bury chemicals very deep in the ground.

    b.   to find a way to make fewer products.

    c.   to make the dumping of wastes unlawful.

5. Which one of the following statements is *true*?

    a.   All chemicals are poisonous.

    b.   Scientists called chemists have stopped making chemicals.

    c.   Chemicals are an important part of our lives.

Trash—A New Energy Source

In some places, trash is being used as fuel.

People throw out trash, or garbage, every day and never think about it. There is so much trash that it has become a problem. We are running out of places to put the trash. Now, scientists have discovered a way to make good use of trash.

In certain parts of the country, trash is collected and brought to special *processing plants* (prŏs′ĕs′ĭng plănts′). At the processing plant, machines remove the glass and metal from the trash. This material is sold. Then the remaining trash is cut up into tiny pieces and put into huge furnaces, where it is burned as fuel.

The burning trash heats water until it becomes steam, and the steam energy drives machines to produce electricity. Finally, the electricity produced at the processing plant is sold to factories or to power companies. Sometimes the steam is also sold for heating purposes.

1. At some *processing plants*, trash is burned as

    a. steam.

    b. fuel.

    c. electricity.

2. What does the processing plant sell besides electricity?

    a. machines and furnaces

    b. glass, metal, and steam

    c. plants and trash

3. Besides trash, what is used in the process described to make electricity?

    a. glass

    b. water

    c. metal

4. Why is trash a good source of fuel?

    a. There is a large supply of it.

    b. It contains glass and metal.

    c. It can be cut up into tiny pieces.

5. Under which of the following headings would you list trash?

    a. Useless Waste Materials

    b. A Source of Energy

    c. An Impossible Problem

Glacier Warning

What is happening to the world's glaciers?

Will glaciers continue to shrink and disappear all together? Many scientists think it is possible. One reason is that the warming temperatures on Earth over the past sixty years have caused the glaciers to melt.

A glacier is made from layers of snow and ice. The snow and ice build up year after year. Glaciers move slowly by the force of gravity. Some can move about 3 feet a day. A glacier is usually long and narrow. Some glaciers are more than 50 miles long and 1,000 feet deep. Glaciers are found in mountainous areas, such as the Alps in Switzerland.

The glaciers of Glacier National Park in Montana are disappearing. In 1850 there were about 150 glaciers in the park. In 2000 there were about 50 glaciers left. Some of the park's glaciers are half of the size they were in 1850. Park scientists report there may not be a single glacier left

in Glacier National Park by the mid-century. Glaciers in the Andes mountains in Peru and glaciers in the Alps are shrinking, too. They are becoming smaller and smaller, year after year.

More than twenty countries have built special glacier stations so scientists can study glaciers. The scientists want to know if the warming conditions on Earth will continue to melt glaciers during this century.

1. Most glaciers are located in

    a. cold mountain areas.

    b. cold oceans.

    c. cold flat places.

2. About how many glaciers disappeared in Glacier National Park from 1850 to 2000?

    a. 50 glaciers

    b. 100 glaciers

    c. 150 glaciers

3. Glaciers move by the force of

    a. wind.

    b. water.

    c. gravity.

4. The main purpose of this story is to tell the reader

    a. that Earth is getting warmer.

    b. how glaciers are made.

    c. that glaciers are disappearing.

5. The Alps are located in

    a. the United States.

    b. Peru.

    c. Switzerland.

Rhinos Need Protection

Something has to be done to save the rhinos.

Environmentalists are working hard to save the African and Asian rhinos. The five kinds include the black, the white, the Indian, the Javan, and the Sumatran. All of them are endangered animals. They are in danger of disappearing from Earth.

Rhinos grow to a shoulder height of 5 feet. They can weigh more than 4,000 pounds. Rhinos eat plants and other vegetation and live in grasslands, open plains, and woody areas. They live about 35 to 40 years. The rhino has one or two horns that grow atop the animal's snout.

In recent decades, rhinos have been hunted nearly to *extinction* by poachers. A poacher is someone who hunts animals in a no-hunting area. Poaching is against the law. The poacher sell the animals' horns. The horns are used for medicine and making ornaments.

Since 1970 the world rhino population has declined by 90 percent. As an example, there are only 2,500 black

rhinos. To save the rhinos, many of the animals have been moved to special parks. The parks protect the animals from poachers. In 1994, for the first time in twenty years, rhino numbers did not decline. The rhinos are still not out of danger. More work is needed to increase the rhinos' numbers.

1. What kinds of foods do rhinos eat?

    a.   small animals

    b.   plants

    c.   plants and animals

2. The word *extinction* in the third paragraph means

    a.   no longer living.

    b.   disappearing in numbers.

    c.   increasing in numbers.

3. The rhinos are disappearing because of

    a.   poaching.

    b.   lack of food.

    c.   attacks by other animals.

4. Rhinos can live to the age of

    a.   15 years.

    b.   35 years.

    c.   55 years.

5. How do parks help save the rhinos?

    a.   They provide food and water.

    b.   They protect the animals from poachers.

    c.   They protect the rhinos from other animals.

Aquaculture, Another Way to Raise Fish

Are there other sources of raising fish and shellfish?

One of the main food industries that will grow in the twenty-first century is aquaculture. Aquaculture is the "farming" of fish, shellfish, or aquatic plants in freshwater or saltwater ponds or lagoons. Fish farms provide much of the trout, catfish, and shellfish eaten in the United States.

There are different kinds of aquaculture. The raising of pet fish for aquariums is one kind. Raising bait fish for the fishing boats is another kind. The most common kind of aquaculture is fish farming and shellfish farming. Fish farming includes raising of fish, such as catfish, trout, and salmon. Shellfish farming includes raising of mussels and shrimp.

Raising fish for food probably began as early as 4,000 years ago in China. Today, fish farming is an important industry in the United States, Philippines, Japan, China, India, Israel, and Europe.

Fish farming may help in the world's food supply. The amount of farm land is becoming less and less. But since most of Earth is covered with water, many people believe that large numbers of fish can be raised for food.

However, there are problems with some fish farms. One problem includes cutting down and destroying marshes and mangroves to build fish farms. The water in the ponds and lagoons can also become polluted.

1. The raising of fish, shellfish and aquatic plants in fresh water or saltwater ponds or lagoons is called _____.

2. Aquaculture provides
   a.   shrimp, tuna, and trout.
   b.   mussels, tuna, and swordfish.
   c.   shrimp, salmon, and catfish.

3. What is one problem of fish farms?
   a.   They cost too much to build.
   b.   They provide only shellfish.
   c.   They destroy marshes.

4. Fish and shellfish farming are important bacuase
   a.   most of the Earth is covered with water.
   b.   they may help the world's food supply.
   c.   they will help stop pollution.

5. Paragraph 3 leads you to believe that
   a.   aquaculture has been around for a long time.
   b.   only three countries use aquaculture.
   c.   aquaculture supplies most of the world's fish.

Don't Bag It. Compost It.

Is there another way of recycling yard wastes?

Recycling wastes is a good way to save Earth's resources. Collecting waste for reuse is called recycling. Examples include collecting bottles and aluminum cans to be used again.

Composting is another way to recycle wastes. Compost is made from wastes such as garbage, grass clippings, twigs, dead leaves, and plants. The wastes are piled up in special containers. The wastes are first broken down by bacteria. Soon fungi and other organisms join the bacteria. Then springtails, mites, and other small insects, as well as earthworms, also help to break down the wastes.

The compost is done when the wastes are changed into a dark soil. Gardeners and farmers can use the dark soil as a natural fertilizer to enrich their soil.

In many cities, trucks collect yard waste from homes and take it to composting plants. In the plant the material is piled into rows, between 3 to 9 feet high. About once a month, machines are used to mix up the piles to add oxygen. Bacteria and other organisms need the oxygen in the air to help break down the wastes. Turning and mixing the pile adds oxygen to the compost. When the compost is done, it is sold or given away. Composting plants are located in the United States, Western Europe, Israel, and Japan.

1. *Recycle* is the word used to describe how waste material is _____ again.

2. A compost pile would most likely contain
    a. newspapers, fungi, and plastic wrappers.
    b. grass clippings, leaves, and earthworms.
    c. plastic wrappers, insects, and paper bags.

3. The wastes in a compost pile are first broken down by
    a. fungi.
    b. bacteria.
    c. earthworms.

4. Oxygen is added to the compost pile because
    a. bacteria needs oxygen to breakdown the wastes
    b. oxygen provides food for living things.
    c. oxygen makes the compost pile darker.

5. The *main* idea of this story is
    a. composting is another way to recycle wastes
    b. recycling wastes is a good way to save Earth's resources
    c. gardeners and farmers use compost to enrich their soil.

The Car of Tomorrow

The car of tomorrow runs on electric batteries.

In the early 1900s, electric cars were quite popular. They ran quietly and smoothly and did not dirty the air with *pollutants* (pə lo͞ot′ nts). But an electric car was slow, and its batteries were very heavy and had to be charged about every 90 miles. Soon, the number of electric cars decreased, and gasoline-burning cars took over the roadways.

The electric car is making a comeback. It still offers a smooth, quiet ride. And it does not give off the pollutants that gasoline-burning cars do. Scientists and engineers have developed smaller and lighter storage batteries. They are working on a battery that will supply even more electricity to the car's motor, so that the car will run at higher speeds.

Some new cars use electric batteries *and* a small gasoline-burning engine. These cars get very good gas mileage and still give off less pollution than gasoline burning cars. In time, maybe the gasoline-burning car may go the way of the covered wagon!

1. A *pollutant* is something that

    a.   dirties the air.

    b.   helps a car ride smoothly.

    c.   makes a battery quiet.

2. The batteries in the electric cars from the early 1900s were very

    a.   quiet.

    b.   heavy.

    c.   small.

3. If a battery that supplies more electricity to the car's motor is developed, then

    a.   electric cars will travel at greater speeds.

    b.   less air-polluting gases will be given off.

    c.   the ride will be much smoother.

4. According to the story, some of the new cars will

    a.   run on two kinds of power.

    b.   use gasoline as a fuel.

    c.   need no motor at all.

5. The last sentence in the story tells us that a gasoline-burning car

    a.   will never replace the covered wagon.

    b.   may soon be a thing of the past.

    c.   will soon be the car of the future.

A Water Power for Energy

How is water used to provide energy for electricity?

Hydroelectric power produces about 22 percent of the world's electricity. Hydroelectric power uses flowing water to drive turbines that generate electricity. Some of the largest hydroelectric power producers are Canada, the United States, Brazil, Norway, Russia, and China. About 10 percent of all the U.S. electricity is produced by water power.

Hydroelectric power is nonpolluting. It produces no harmful air pollutants and no liquid or solid wastes. However, environmentalists state that building large dams for water power harms the environment. Building a dam means flooding a large area behind the dam. The flooding of the land makes a large lake. The water in the lake is used by the plant. But the flooding also destroys wetlands and farmlands where plants and animals live.

Building smaller, rather than larger, hydroelectric power plants may be better. The small hydroelectric plants can be built without damming rivers. They are built on rivers but do not stop the flow of the water. They do not harm the environment as much as the larger plants. They are less costly to build and maintain than the larger plants. These hydroelectric plants can be used in small villages and towns to generate electricity for lighting and for pumping water. Small hydroelectric plants are used in China, United States, Indonesia, Nepal, Sri Lanka, and Zaire.

1. According to the story, which of the following statements is true. Hydroelectric power produces

    a.   about 25 percent of all electricity in the United States.

    b.   less than 5 percent of all electricity in the United States.

    c.   about 10 percent of all electricity in the United States.

2. Hydroelectric plants are built to produce

    a.   electricity.

    b.   heat.

    c.   light.

3. The major hydroelectric power producers include

    a.   China, Norway, and Sweden.

    b.   China, Brazil, and Italy.

    c.   China, United States, and Brazil.

4. The water behind a dam of a hydroelectric plant is used

    a.   for fishing and boating activities.

    b.   to drive turbines for electricity.

    c.   for drinking water.

5. How are large hydroelectric plants and small hydroelectric plants alike?

    a.   They need large dams.

    b.   They produce electricity.

    c.   They cost the same to build.

BIBLIOGRAPHY

Books on Life Science

_____. *Human Body*. New York: Time-Life, 1992.

_____. *Project Panda Watch*. New York: Macmillan, 1984.

Boulton, Carolyn. *Trees*. New York: F. Watts, 1984.

Casselli, Giovanni. *The Human Boday and How It Works*. New York: Putnam, 1987.

Florian, Douglass. *Discovering Trees*. New York: Scribners, 1986.

Henderson, Douglass. *Dinosaur Tree*. New York: Simon & Schuster, 1994.

Lorimer, Lawrence T.; Fowler, Keith. *The Human Body: A Fascinating See-Through View of How Bodies Work*, Pleasantville, NY: Readers Digest, 1999

Maestro, Betsy. *How Do Apples Grow?*. New York: Harper Collins, 1992.

Owen, Jennifer. *Insect Life*. Tulsa, Oklahoma: EDC Publications, 1985.

Showers, Paul. *How Many Teeth?*. New York: Harper Collins, 1991.

Silverstein, Alvin; Silverstein, Virginia; Silverstein, Laura. *Common Colds (My Health)*. New York: F. Watts, 1999.

Stockley, Corinne. *Animal Behavior, Science and Nature Series*. Tulsa, OK: EDC Publications, 1992.

Waldbauer, Gilbert. *Millionsof Monarchs, Bunches of Beetles: How Bugs Find Strength in Numbers*. Cambridge, MA: Harvard Univ. Press, 2000

Books on Earth-Space Science

_____. *Iceburgs & Glaciers*. New York: Morrow, 1987.

_____. *Space Planets*, New York: Time-Life, 1992.

Blair, Carvel. *Exploring the Sea: Oceanography Today*. New York: Random House, 1986.

Bramwell, Martyn. *Planet Earth*. New York: F. Watts, 1987.

Bramwell, Martyn. *Rocks and Fossils*. Tulsa, OK: E D C Publishing, 1994.

Bramwell, Martyn. *The Oceans*. New York: F. Watts, 1994.

Bramwell, Martyn. *Volcanoes and Earthquakes*. New York: F. Watts, 1994.

Branley, Franklin M. *Sunshine Makes the Seasons*. rev. ed. New York: Crowell, 1986.

Branley, Franklin M. *The Sun, Our Nearest Star*. New York: Crowell, 1988.

Branley, Franklin. *Floating in Space*. New York: Harper Collins, 1998.

Cottonwood, Joe. *Quake!*. New York: Apple, 1996.

Lauber, Patricia. *You're Aboard Spaceship Earth*. New York: Harper Collins, 1996.

Morris, Neil. *Earthquakes* (Wonders of Our World, No. 1). New York, NY: Crabtree Publishing, 1998.

Pollard, Michael. *Air, Water, Weather*. New York: Facts on File, 1987.

Ricciuti, Edward R.; *Carruthers*, Margaret W. Rocks and Minerals (National Audubon Society). New York: Scholastic, Inc., 1998.

Simon, Seymour. *Danger from Below-Earthquakes: Past, Present, and Future*. New York: Four Winds Press, 1979.

Books on Physical Science

_____. *Physical Forces*. New York: Time-Life, 1992.

Ardley, Neil. *Making Things Move*. New York: F. Watts, 1984.

Barrett, Norman. *Sports Machines*. New York: F. Watts, 1994.

Bramwell, Martyn. *Glaciers and Ice Caps*. New York: F. Watts, 1994.

Branley, Franklin M. *Gravity Is a Mystery*. rev. ed. New York: Crowell, 1986.

Branley, Franklin Mansfield. *The Beginning of the Earth*. New York: Harper Row, 1988.

Durant, Penny Raife. *Make a Splash*. New York: F. Watts, 1991.

Haslam, Andrew; Glover, David. *Machines (Make It Work! Series)*. Pittsfield, MA: World Book, Inc., 1997.

Morgan, Nina. *Lasers* (20th Century Inventions). Texas: Raintree Steck-Vaughn, 1997.

Walpole, Brenda. *Movement, Fun with Science Series*. New York: Gloucester Press, 1987.

Whyman, Kathryn. *Heat and Energy*. New York: Gloucester Press, 1986.

Books on Environmental Science

Accorsi, William. *Rachel Carson*. New York, NY: Holiday House, 1993.

BIBLIOGRAPHY

Berger, Melvin. *Oil Spill!*, New York: Harper Collins, 1994.

Forman, Michael H. *Arctic Tundra* (Habitats). Danbury, CT: Children's Press, 1997.

George, Jean Craighead. *1 Day in the Tropical Rain Forest* (Newbery Medal Winner Series, No. 5). New York: Crowell Co., 1990.

Mongillo, John; Zierdt-Warshaw, Linda. *The Encyclopedia of Environmental Science*. Phoenix, AZ: Oryx Press, 2001

Morgan, Sally. *Acid Rain* (Earth Watch). New York: F. Watts, 1999.

Pollock, Steve. *The Atlas of Endangered Animals* (Environmental Atlas Series). New York: Checkmark Books, 1993.

RECORD KEEPING

The Progress Charts on these pages are for use with questions that follow the stories in the Life Science, Earth-Space Science, Physical Science, and Environmental Science Units. Keeping a record of your progress will help you see how well you are doing and where you need to improve. Use the charts in the following way:

After you have checked your answers, look at the first column, headed "Questions Page." Read down the column until you find the row with the page number of the questions you have completed. Put an X through the number of each question in the row that you have answered correctly. Add the number of correct answers, and write your total score in the last column in that row.

After you have done the questions for several stories, check to see which questions you answered correctly. Which ones were incorrect? Is there a pattern? For example, you may find that you have answered most of the literal comprehension questions correctly but that you are having difficulty answering the applied comprehension questions. If so, then it is an area in which you need help.

When you have completed all of the stories in an unit, write the total number of correct answers at the bottom of each column.

PROGRESS CHART FOR LIFE SCIENCE UNIT

Questions Page	Comprehension Question Numbers				Total Number Correct per Story
	Science Vocabulary	Literal	Interpretive	Applied	
7	1	2	3,4,5		
9	1	2	3,4	5	
11	1	2	3,4,5		
13	1	2	3,4,5		
15	1	2	3,4,5		
17	1,2	3	4,5		
19	1	2	3,4	5	
21	1	2	3,4,5		
23	1	2,3	4	5	
25	1	2	3,4,5		
27	1	2	3,4		
29	1	2,3	4	5	
31	1	2	3,4	5	
33	1	2,3	4,5		
Total Correct by Question Type					

PROGRESS CHART FOR EARTH-SPACE SCIENCE UNIT

Questions Page	Comprehension Question Numbers				Total Number Correct per Story
	Science Vocabulary	Literal	Interpretive	Applied	
39	1	2,3	4,5		
41	1	2,3	4,5		
43	1	2,3	4	5	
45	1	2	3,4,5		
47	1	2,3	4	5	
49		2,3	4,5		
51		1,2,3	4	5	
53	1	2	3,4,5		
55	1	2,3	4,5		
Total Correct by Question Type					

PROGRESS CHART FOR ENVIRONMENTAL SCIENCE UNIT

Questions Page	Comprehension Question Numbers				Total Number Correct per Story
	Science Vocabulary	Literal	Interpretive	Applied	
81	1	2	3,4,5		
83	1	2,3	4,5		
85	1	2	3,4	5	
87		1,2,3	4	5	
89	2	1,3,4,5			
91	1	2,3	4,5		
93	1	2,3,4	5		
95	1	2,3	4,5		
97		1,2,3	4,5		
Total Correct by Question Type					

PROGRESS CHART FOR PHYSICAL SCIENCE UNIT

Questions Page	Comprehension Question Numbers				Total Number Correct per Story
	Science Vocabulary	Literal	Interpretive	Applied	
61		2,3,4	1,5		
63		1,2,3	4,5		
65		1,2,3	4,5		
67	1	2,3	4,5		
69	1	2,3	4,5		
71	1	2,3	4,5		
73	4	1,2,3	5		
75	1	2,3	4,5		
Total Correct by Question Type					

METRIC TABLES

This table tells you how to change customary units of measure to metric units of measure. The answers you get will not be exact.

LENGTH

Symbol	When You Know	Multiply by	To Find	Symbol
in	inches	2.5	centimeters	cm
ft	feet	30	centimeters	cm
yd	yards	0.9	meters	m
mi	miles	1.6	kilometers	km

AREA

Symbol	When You Know	Multiply by	To Find	Symbol
in^2	square inches	6.5	square centimeters	cm^2
ft^2	square feet	0.09	square centimeters	cm^2
yd^2	square yards	0.8	square meters	m^2
mi^2	square miles	2.6	square kilometers	km^2
	acres	0.4	hectares	ha

MASS (WEIGHT)

Symbol	When You Know	Multiply by	To Find	Symbol
oz	ounces	28	grams	g
lb	pounds	0.45	kilograms	kg
	short tons (200 lb)	0.9	tonnes	t

VOLUME

Symbol	When You Know	Multiply by	To Find	Symbol
tsp	teaspoons	5	milliliters	mL
Tbsp	tablespoons	15	milliliters	mL
fl oz	fluid ounces	30	milliliters	mL
c	cups	0.24	liters	L
pt	pints	0.47	liters	L
qt	quarts	0.95	liters	L
gal	gallons	3.8	liters	L
ft^3	cubic feet	0.03	cubic meters	m^3
yd^3	cubic yards	0.76	cubic meters	m^3

TEMPERATURE (exact)

Symbol	When You Know	Multiply by	To Find	Symbol
°F	Fahrenheit temperature	5/9 (after subtracting 32)	Celsius temperature	°C

METRIC TABLES

This table tells you how to change metric units of measure to customary units of measure. The answers you get will not be exact.

LENGTH

Symbol	When You Know	Multiply by	To Find	Symbol
mm	millimeters	0.04	inches	in
cm	centimeters	0.4	inches	in
m	meters	3.3	feet	ft
m	meters	1.1	yards	yd
km	kilometers	0.6	miles	mi

AREA

Symbol	When You Know	Multiply by	To Find	Symbol
cm^2	square centimeters	0.16	square inches	in^2
m^2	square meters	1.2	square yards	yd^2
km^2	square kilometers	0.4	square miles	mi^2
ha	hectares (10,000 m^2)	2.5	acres	

MASS (WEIGHT)

Symbol	When You Know	Multiply by	To Find	Symbol
g	grams	0.035	ounces	oz
kg	kilograms	2.2	pounds	lb
t	tonnes (1000 kg)	1.1	short tons	

VOLUME

Symbol	When You Know	Multiply by	To Find	Symbol
mL	milliliters	0.03	fluid ounces	fl oz
L	liters	2.1	pints	pt
L	liters	1.06	quarts	qt
L	liters	0.26	gallons	gal
m^3	cubic meters	35	cubic feet	ft^3
m^3	cubic meters	1.3	cubic yards	yd^3

TEMPERATURE (exact)

Symbol	When You Know	Multiply by	To Find	Symbol
°C	Celsius temperature	9/5 (then add 32)	Fahrenheit temperature	°F